She Spoke of Her River

By Pam Backlund

May our children and our
children's children
always have
clean water to drink,
and a delightful river to visit.

- Pam Backlund 1998

She Spoke of Her River by Pam Backlund
Published by River Avenue Press
58 West 34th Street, San Angelo, Texas 76903

Cover by Bruce Backlund
Illustrations by Pam Backlund
ISBN: **978-1-7345776-4-8**
LCCN— 2022912194
1. Main Category- History
2. Other Category- Science
Printed in the United States of America
Edition 2

She Spoke of Her River

By Pam Backlund

Table of Contents

Introduction

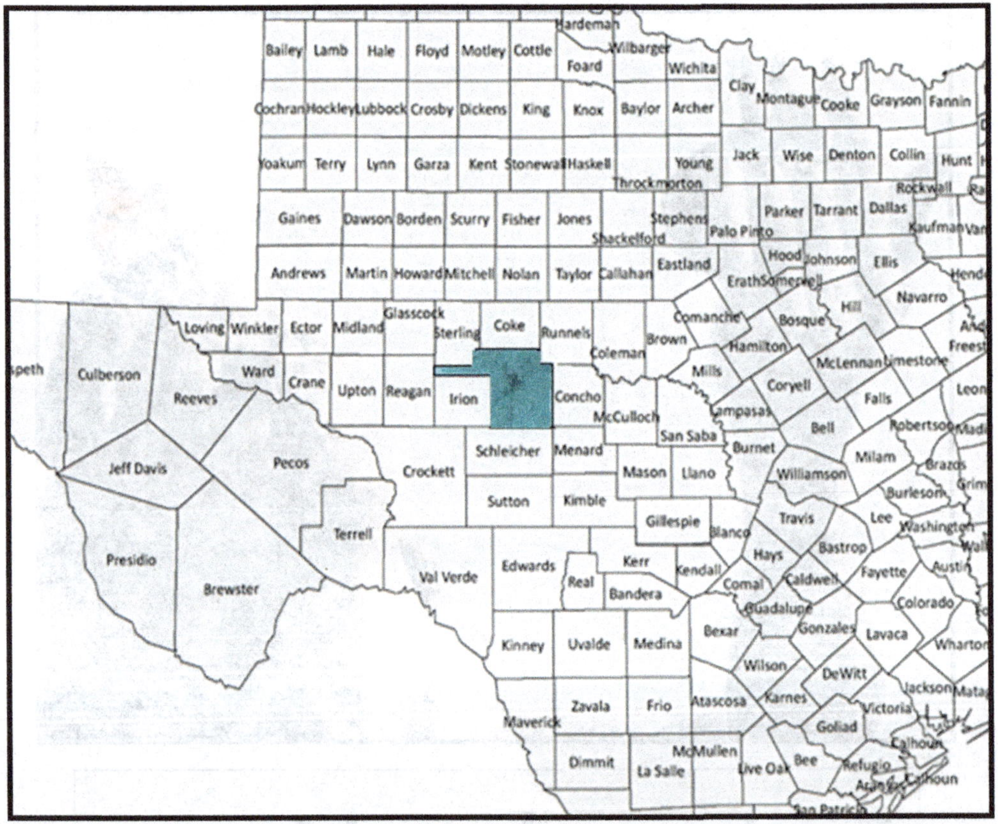

This is a story about the Concho River.

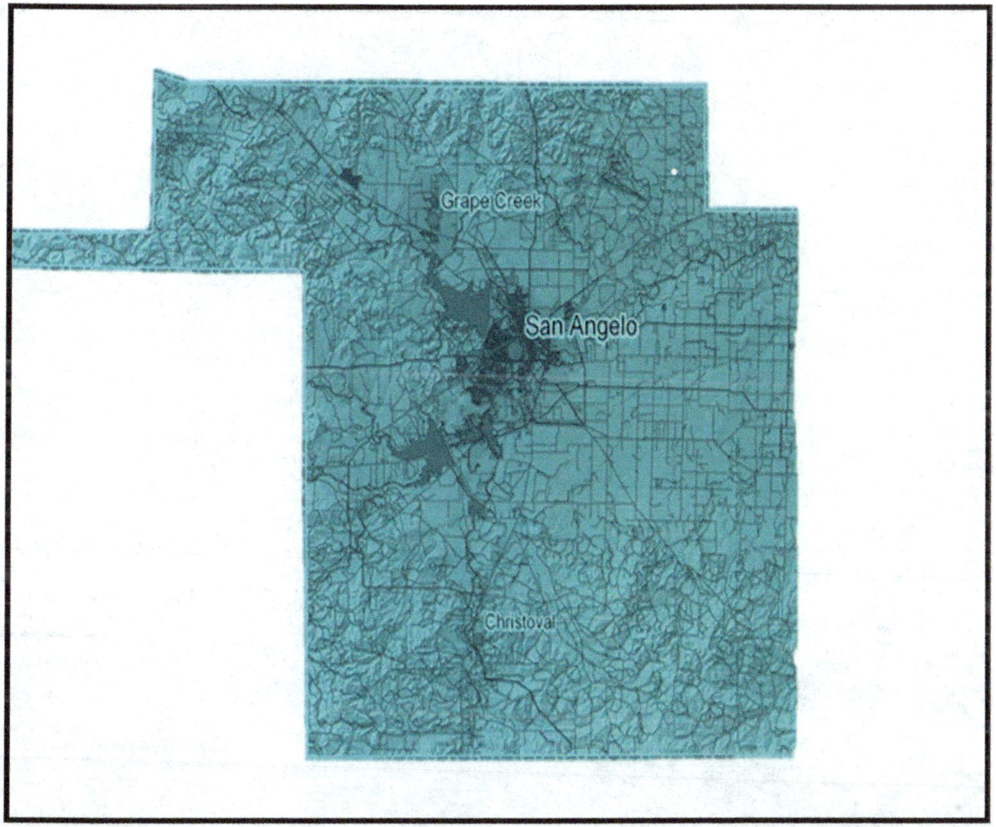

It's a fictional story about a real river in Texas.

In Texas, the city of San Angelo exists.

The city has a Museum of Fine Arts.

The museum is near the Concho River...

...where a mermaid statue sits.

Once a year, Monarch butterflies migrate through the city in massive flocks.

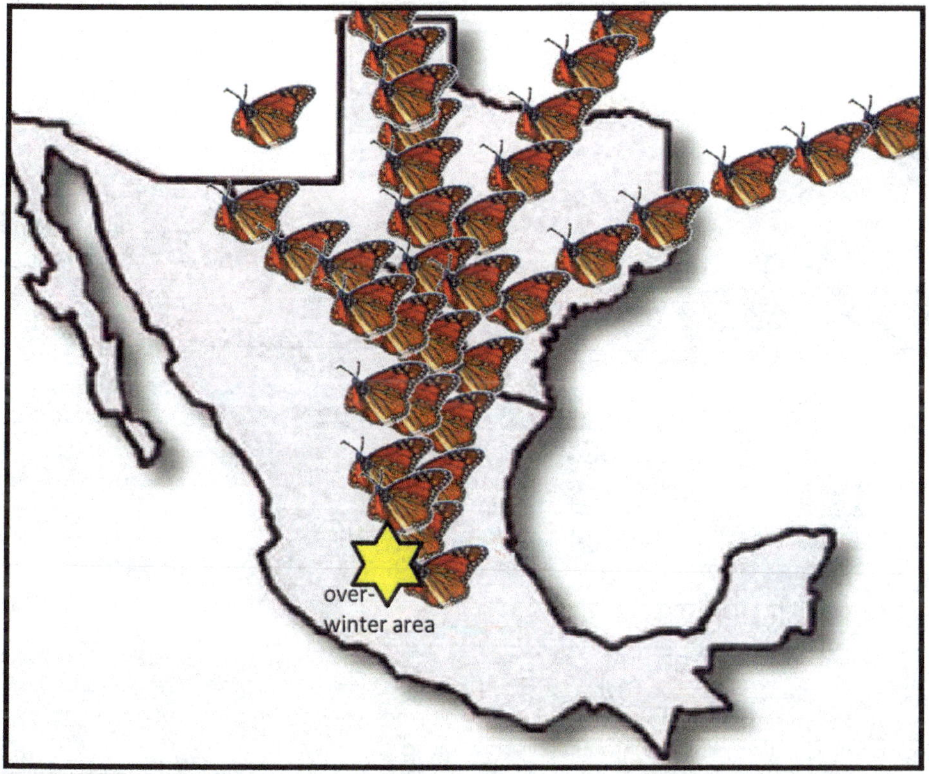

Their journey ends in Mexico,
where they overwinter.

The following story is...

...of course, fiction.

Because migrating butterflies can't connect a mermaid statue with two innocent kids.

And a statue can't be passionate about river history and river protection.

But let's allow the flock of butterflies to start the story, anyway.

Chapter 1
Rain, Rain, Rain

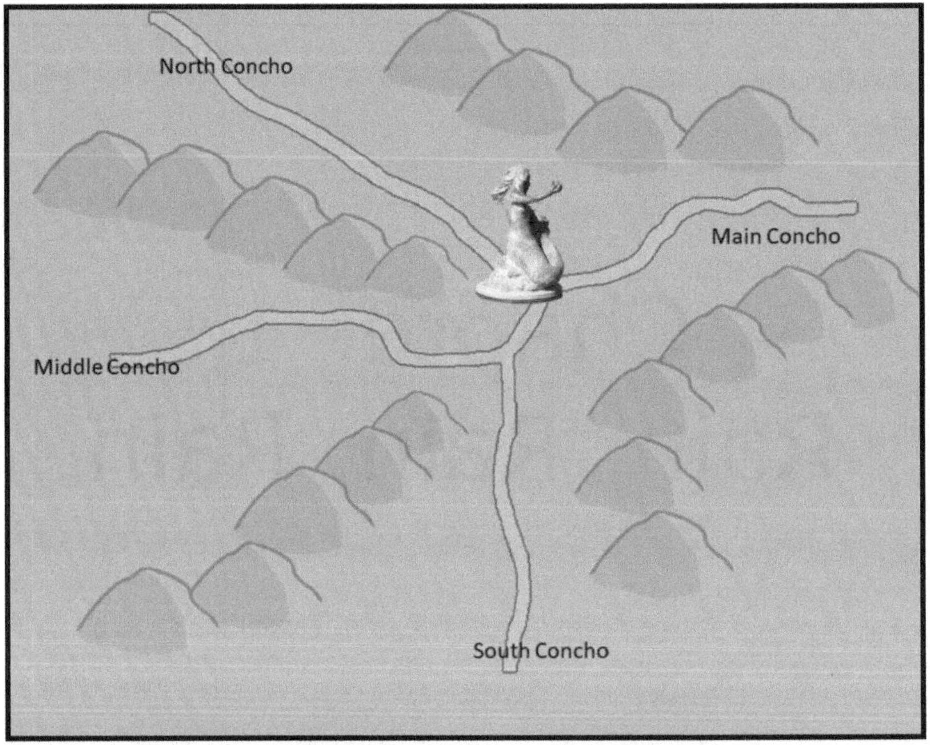

Once there was a river with a mermaid statue in it.

Three kids and a dad were nearby, enjoying the day.

Dezzerae, Miranda, and Haley were chasing a Monarch butterfly which eventually landed on the mermaid statue.

They couldn't catch the butterfly without falling in the water, so they waited and watched.

Miranda’s dad settled down on the Celebration Bridge to read to his daughter.

Suddenly, the mermaid statue began speaking.

Surprisingly she asked, "Has a pipe ever broken at your house, forcing you to have no water for a day?"

The kids looked at each other.

They looked at Miranda and her dad
who was intently reading.
Neither had heard the mermaid speak.

The mermaid asked another question.
"Why do people need river and lake water?"

Moments passed.

Finally, Dezzerae said, "For fishing."

Haley added, “For skiing.”

Dezzerae continued, "For swimming, too."

The mermaid smiled sweetly,
"You are both right."

"But mostly, all living things really need water to survive."

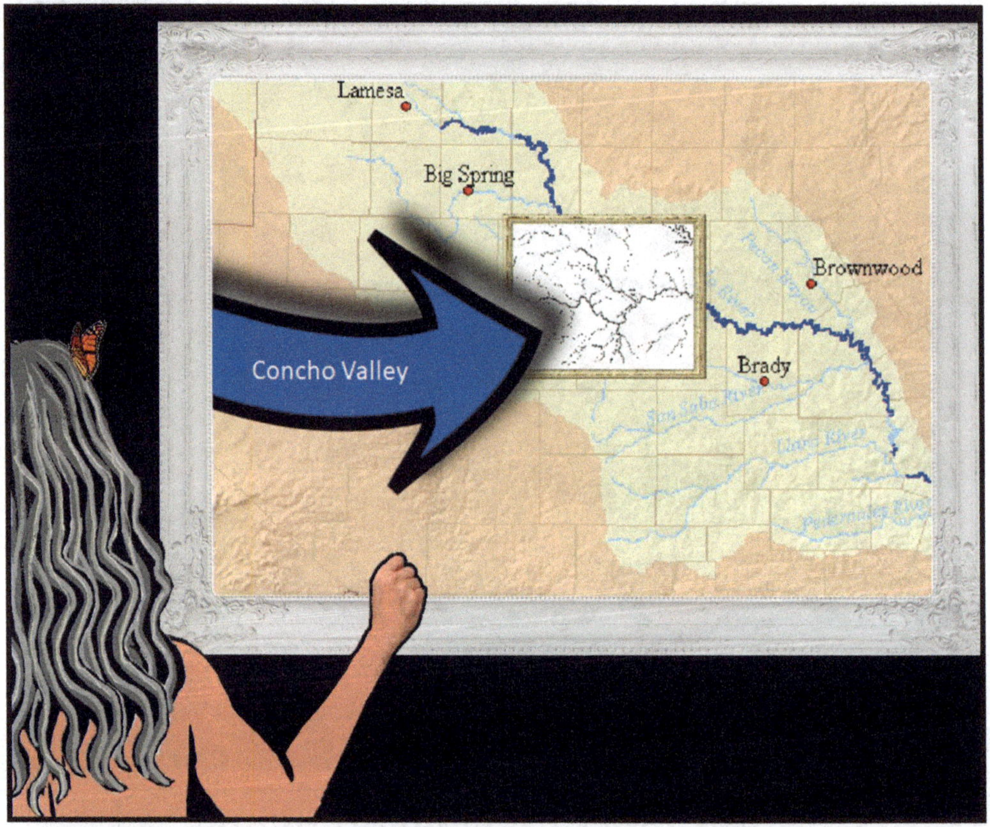

"That is why this river is so special."

During another moment of silence, Haley thought she might not speak again.

"Please keep talking to us,
Ms. Mermaid," he begged.

Three more butterflies landed on the mermaid. She continued.

"Do you children know much about my river?"

Not knowing anything about the river, Dezzerae asked, "What is there to know about a river?"

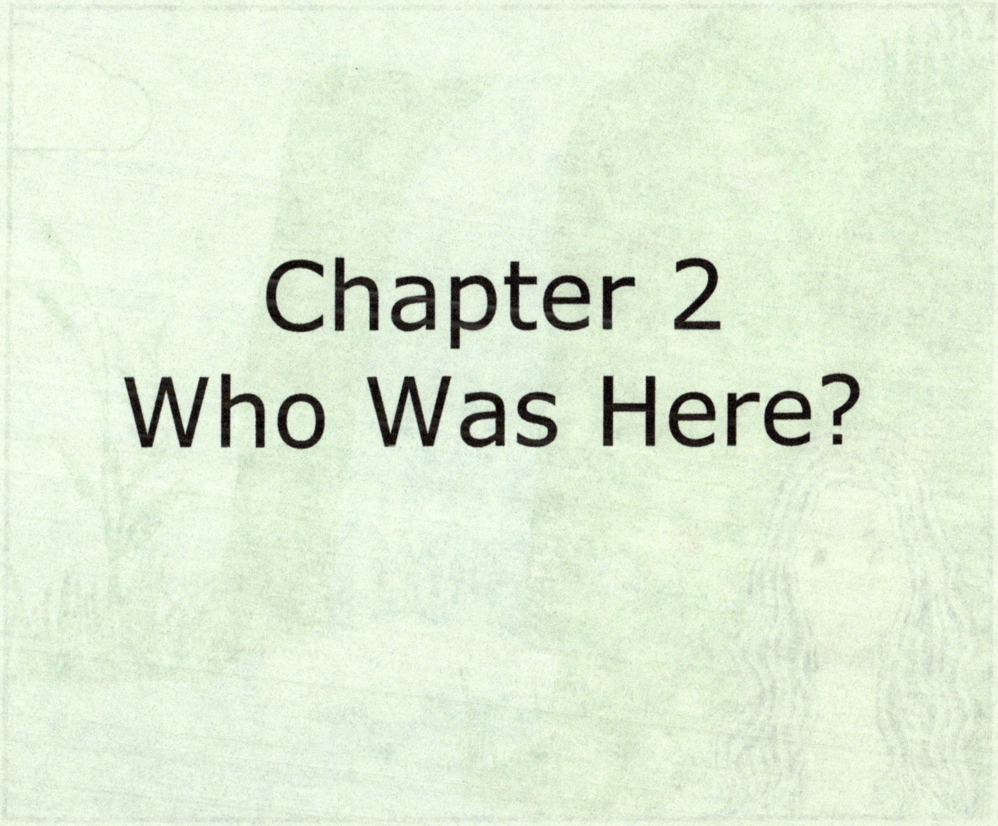

Chapter 2
Who Was Here?

"Well my river has a history I'd like to share."

"A long time ago, the river and its land were home to many wild things..."

"...such as buffalo, butterflies..."

"...birds and bats."

"Grass covered most of the land."

"Most trees grew only directly by the river."

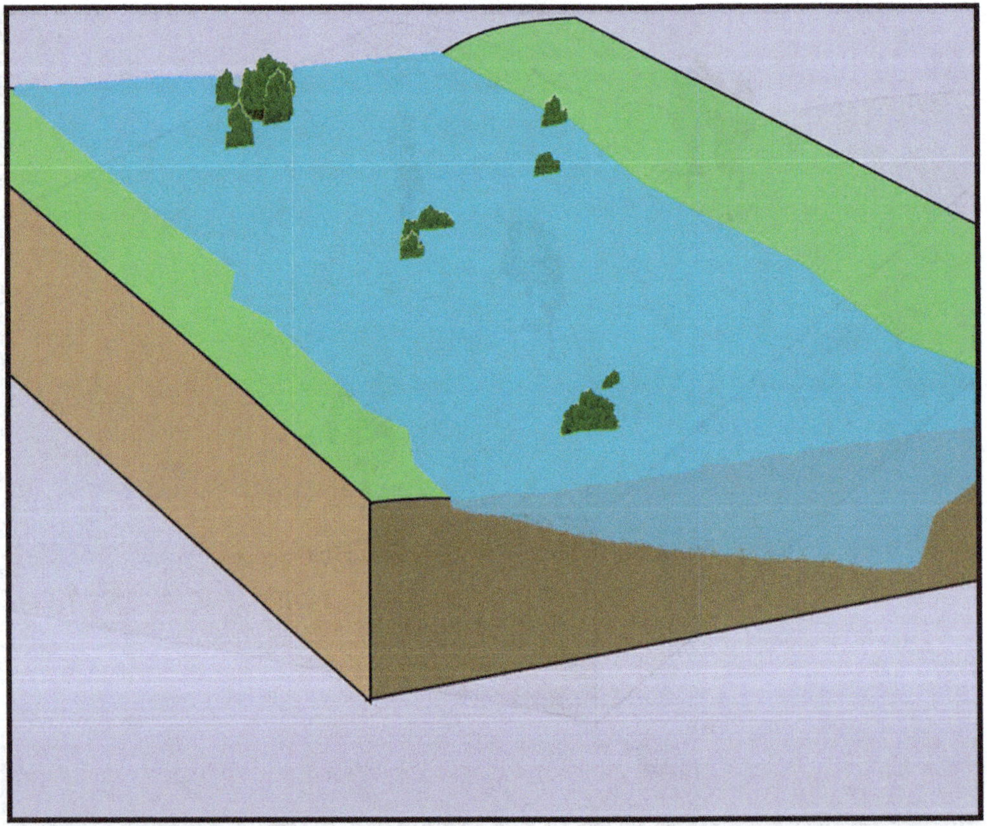

"Sometimes the river was deep and wide."

"Other times the river was shallow and thin."

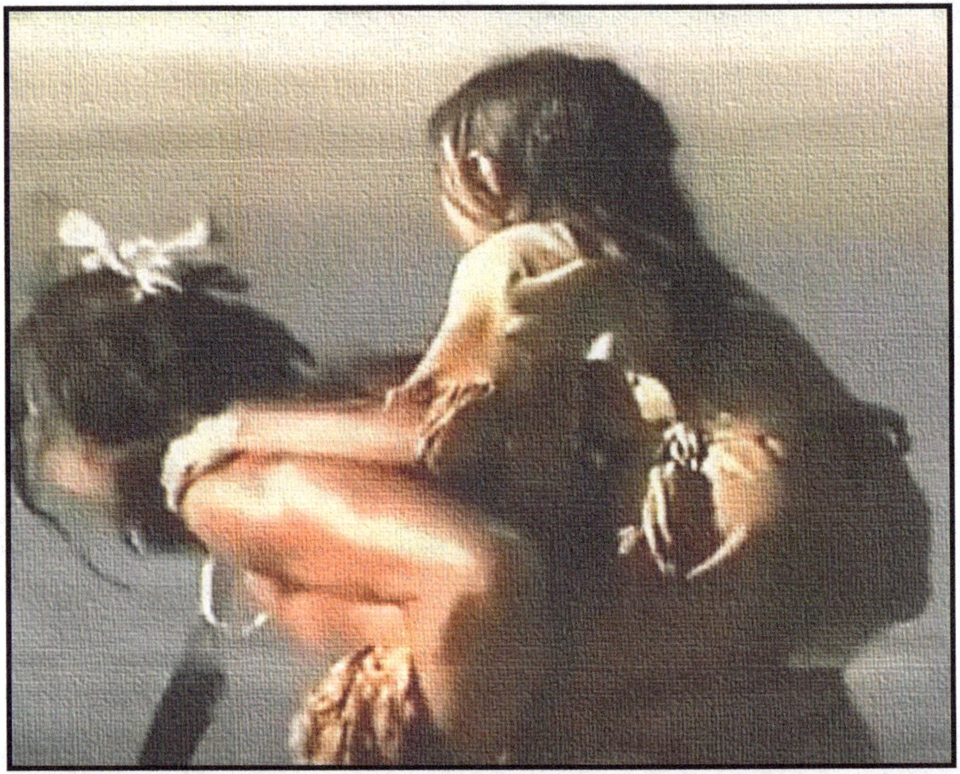

"Native Americans occasionally came to the valley to enjoy the refreshing river and land."

"It was a lonely time for me, because for a long time, only a relatively few individuals depended on this changing river," said the mermaid.

Haley asked, "Why did the river change?"

Image: National Science Foundation

The mermaid shared her wisdom.
"Rivers are always changing.
That's just nature's way."

Normal depth (cross section view) of a typical river.

Normal depth
(top cartoon view.)

"More rain means..."

"...fast-moving, deeper water."

"Less rain..."

"...means slow-moving, shallow water."

Rain changes

cause

river changes.

"Because the weather changes, the river does, too."

"But people can always find beauty in change, if they want to."

"In fact, many people have found goodness in my river."

"Many people have decided,
throughout time, to live here..."

"...and they have been welcome."

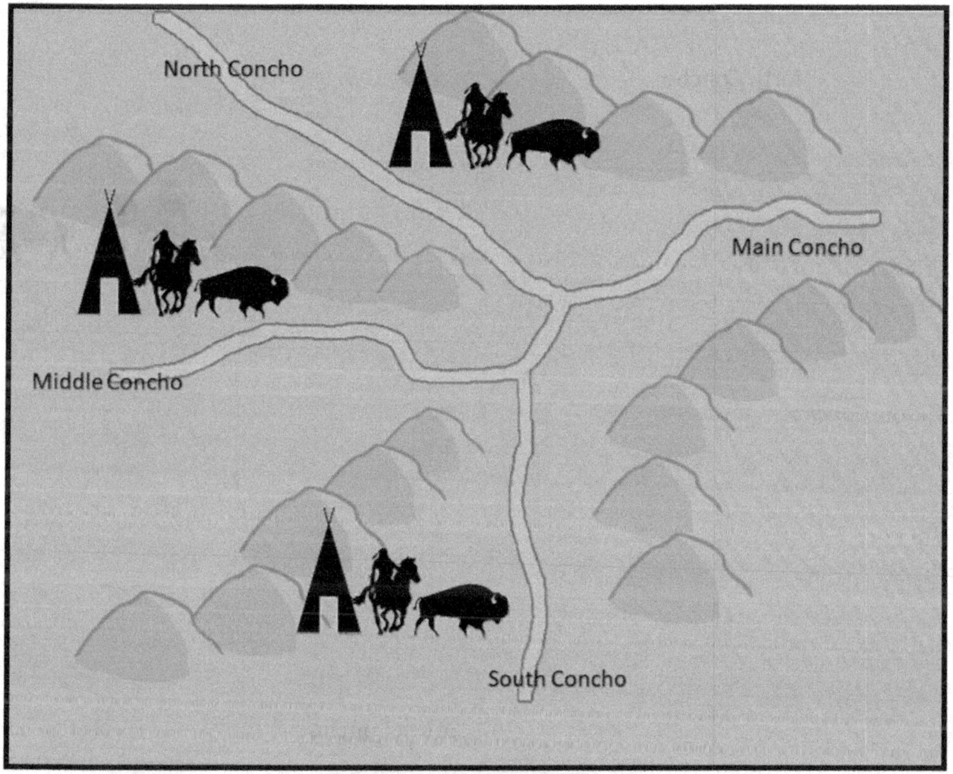

"The first people of my river, the Native Americans, never stayed here very long."

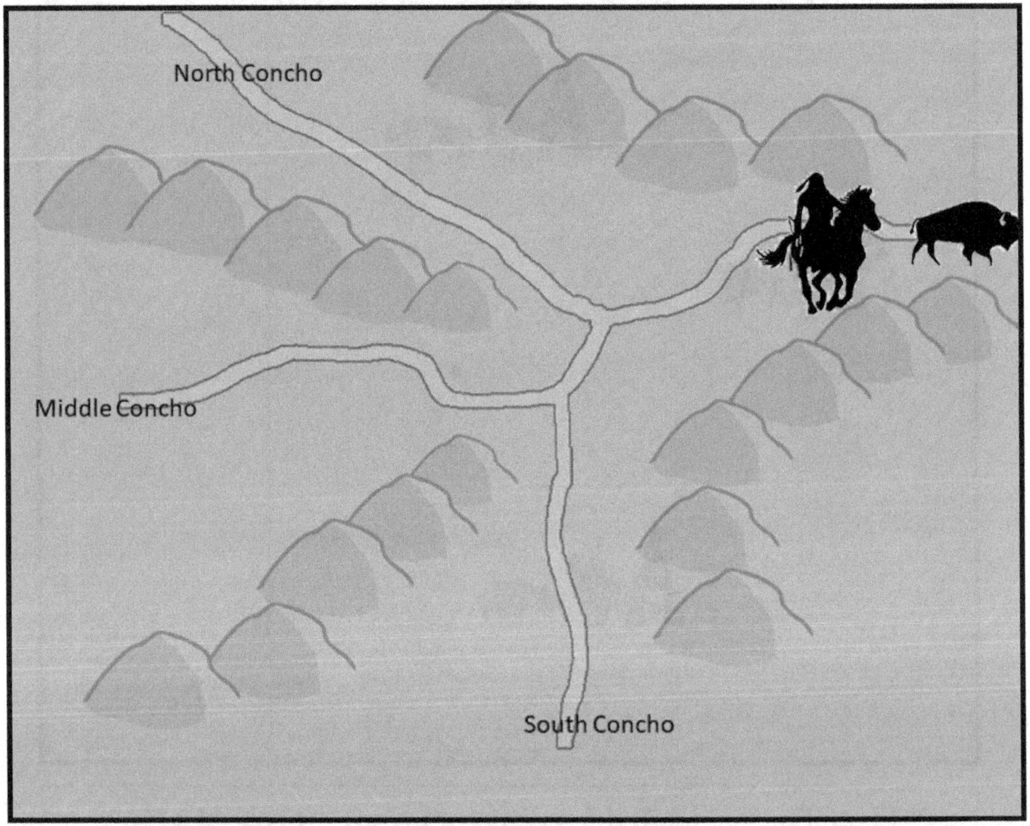

"They migrated with the buffalo."

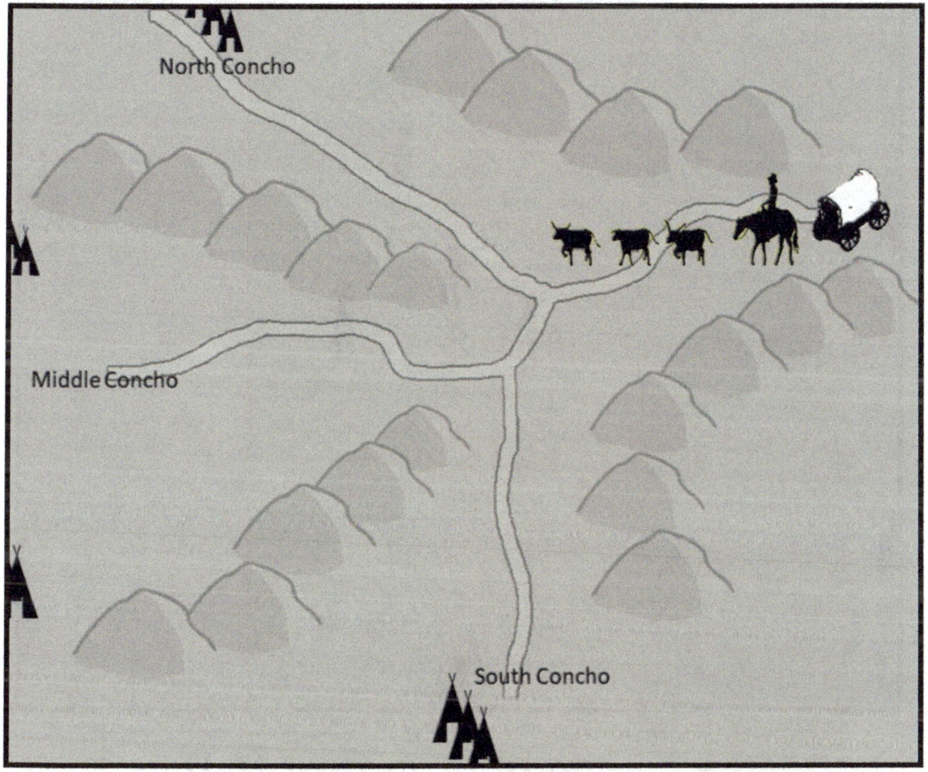

"But when other people wanted to travel through or actually live near my river..."

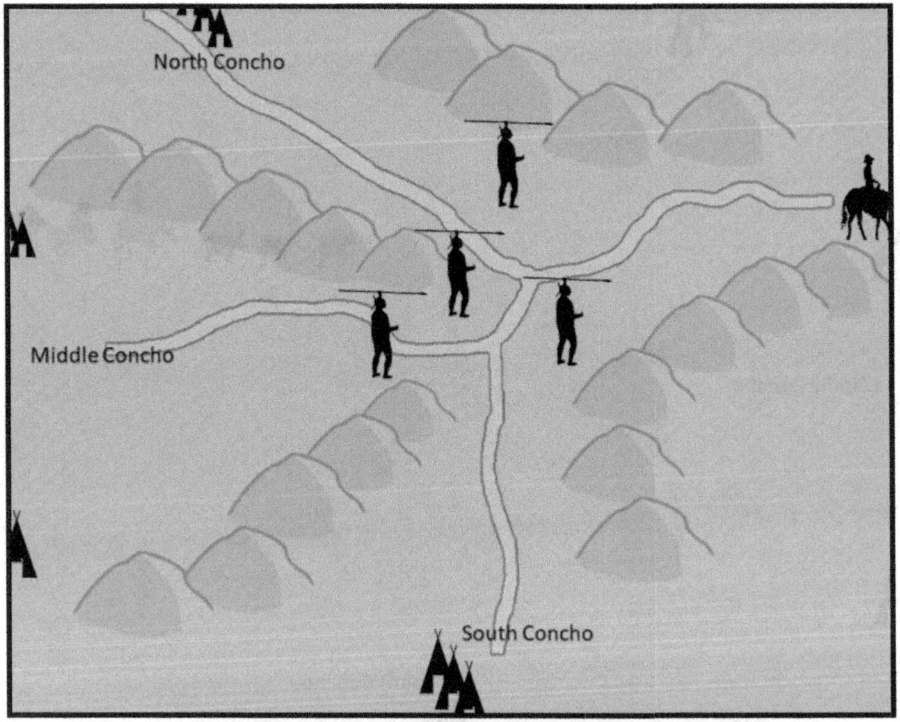

"...a couple of groups of Native Americans wanted to scare the new people away."

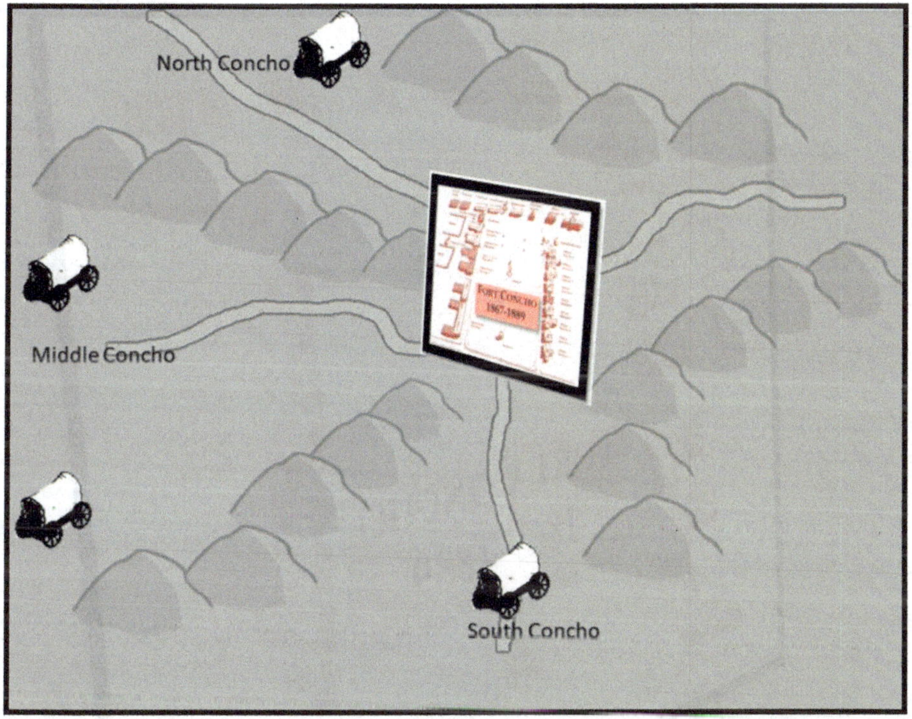

"So, Fort Concho was built to show the Native Americans that new people can live by my river."

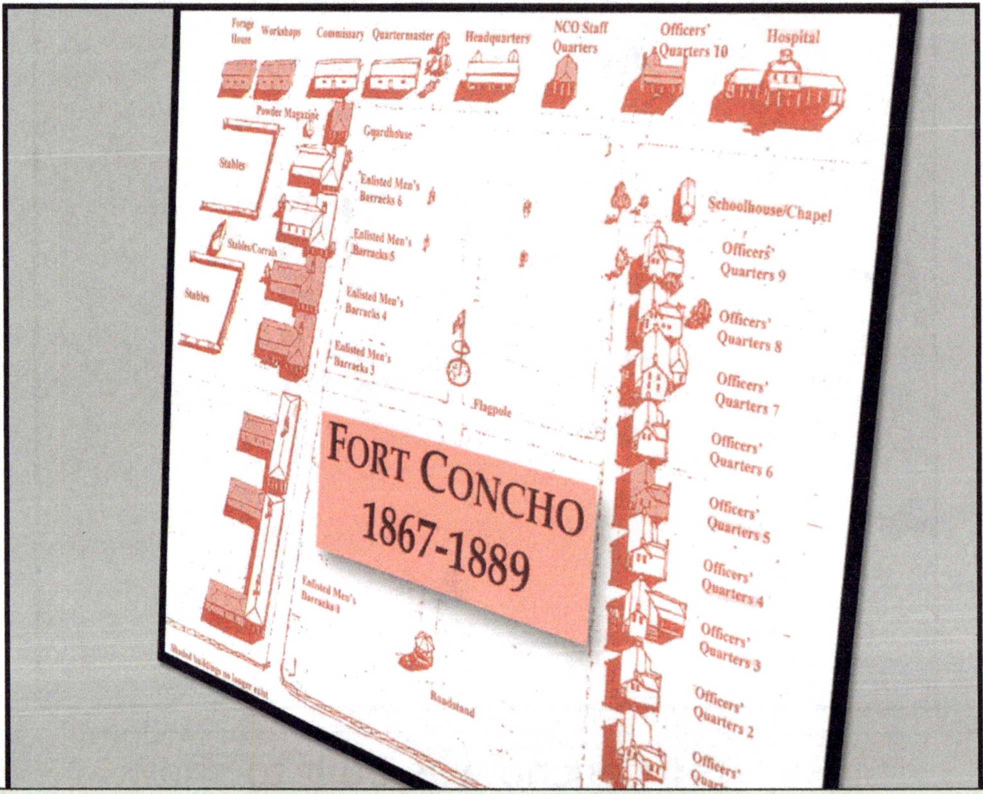

"Various soldiers stayed at the Fort for 22 years."

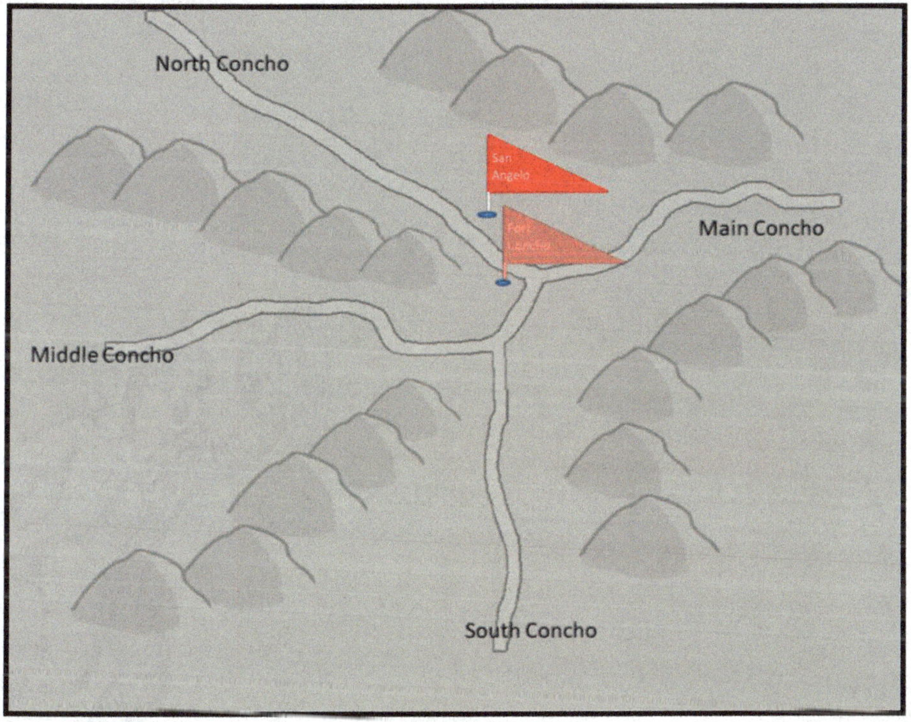

"San Angelo or 'town across the river' became home to many newcomers, partly because of the Fort, and partly because it is a really nice place to live."

"My river was also special ..."

"...to the new people of the 1860s - 1890s."

"As time went on, more people,
more plants, and more animals
began to depend on the river water."

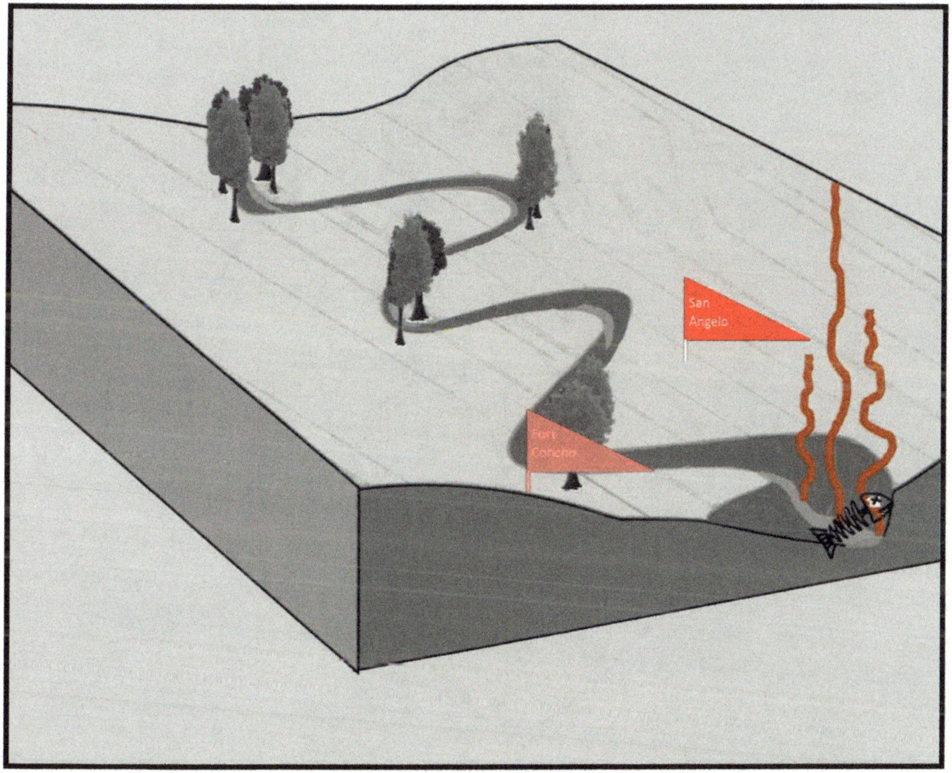

“Sometimes the river became so shallow, and slow-moving and warm, that it began to stink.”

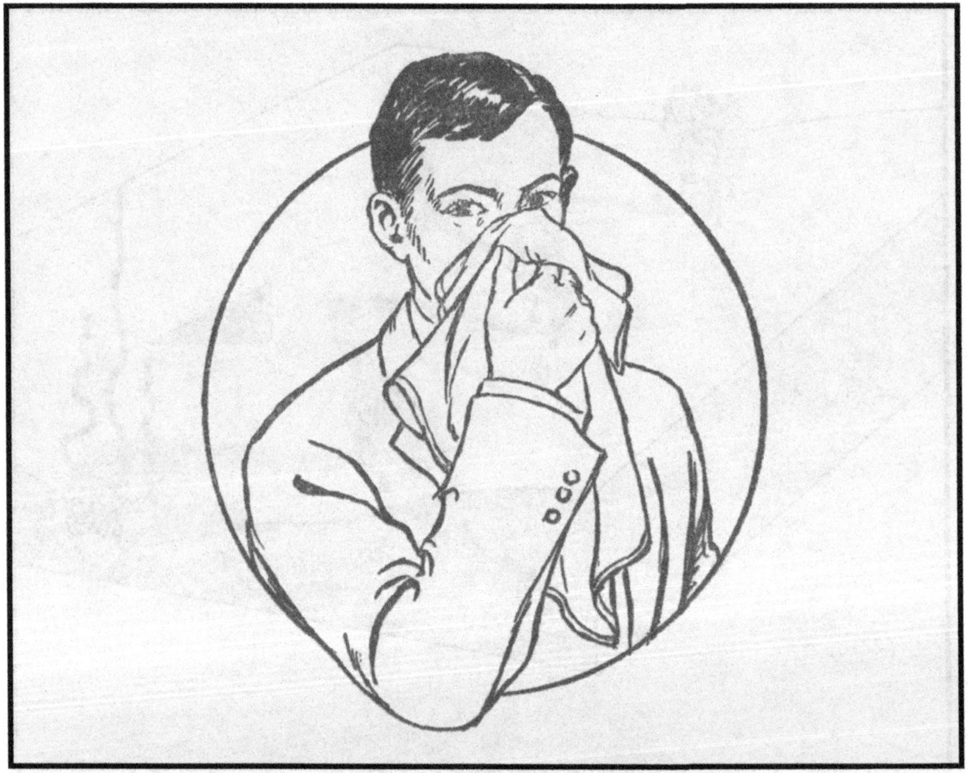

"The smelly river was not a welcome change."

Two more butterflies sat on
the mermaid as she said...

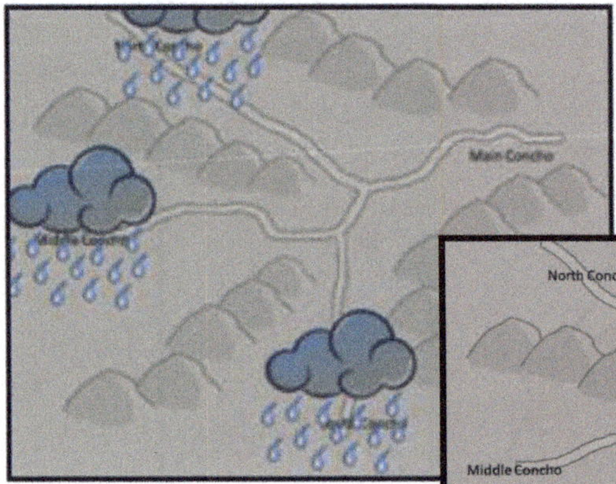

Later in time

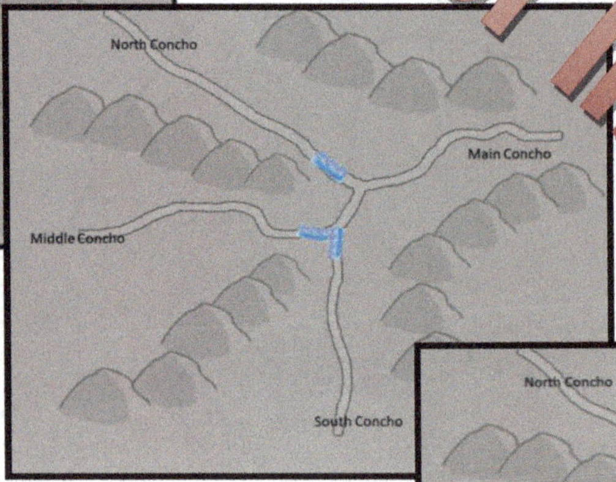

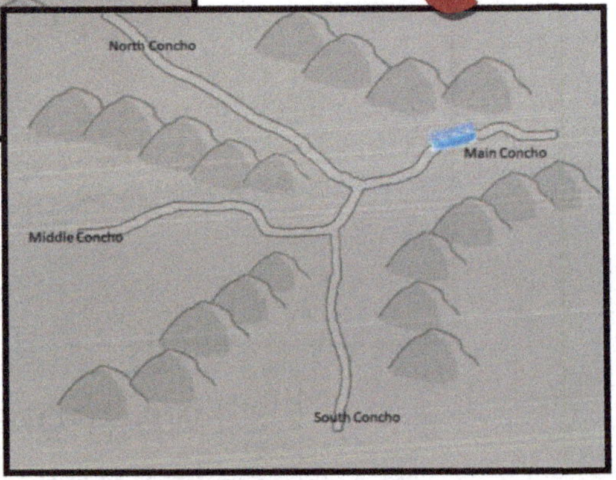

"...Even when the river became stinky though, a big rain would eventually..."

"...come along to wash the stinky stuff away."

"A cleansing rain is a good thing for a river."

"So shallow, warm and slow rivers become stinky," concluded Dezzerae.
"Yes," she answered.

Chapter 3
Using water

A soft breeze fluttered the pages of Miranda's dad's book. He kept reading.

Photo: West Texas Collection, Angelo State University

"Did the people in the past actually drink river water? How could anyone drink stinky water?" Haley asked.

She answered, "People got sick from drinking stinky water."

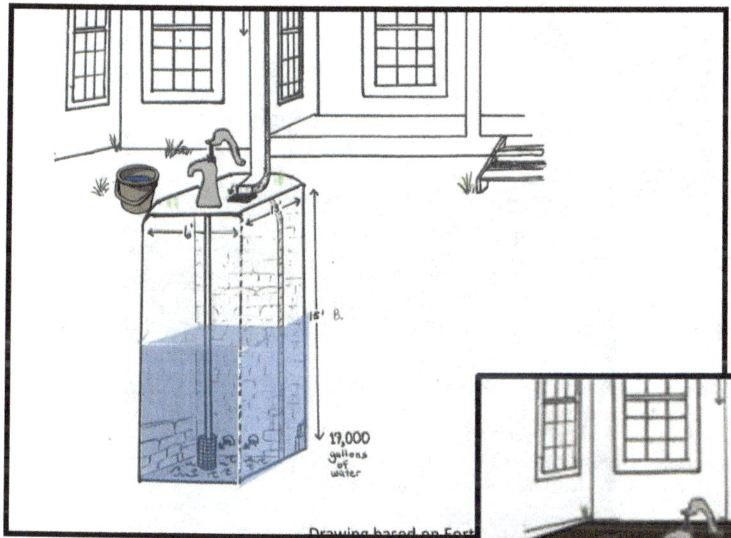

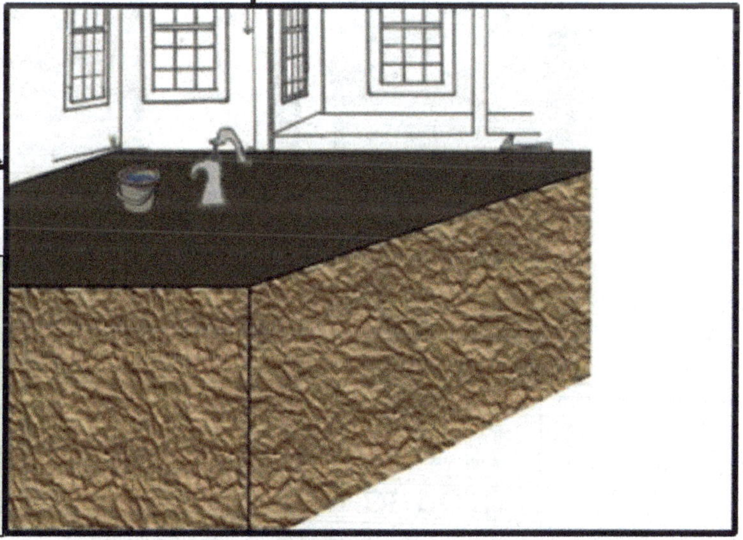

"So they collected rainwater off of their roofs...sometimes storing it underground."

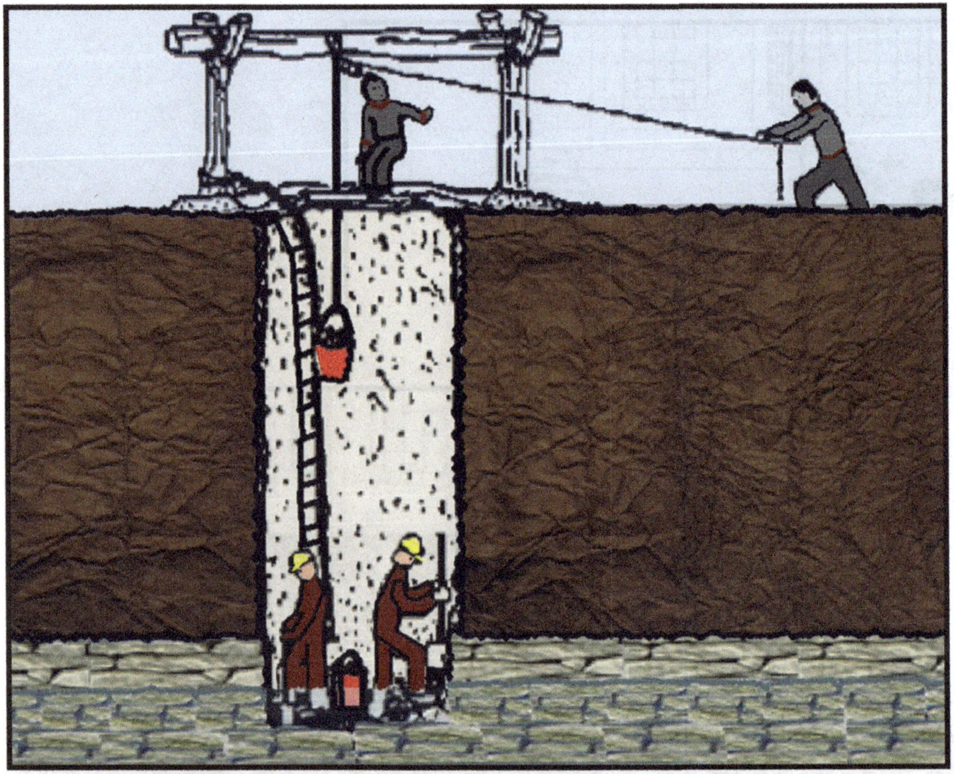

"And they dug deep holes in the ground and found good water there."

"Some put windmills over the holes or wells, to pull the groundwater up for their families, gardens and animals."

"People didn't worry about water as much, once they could get groundwater."

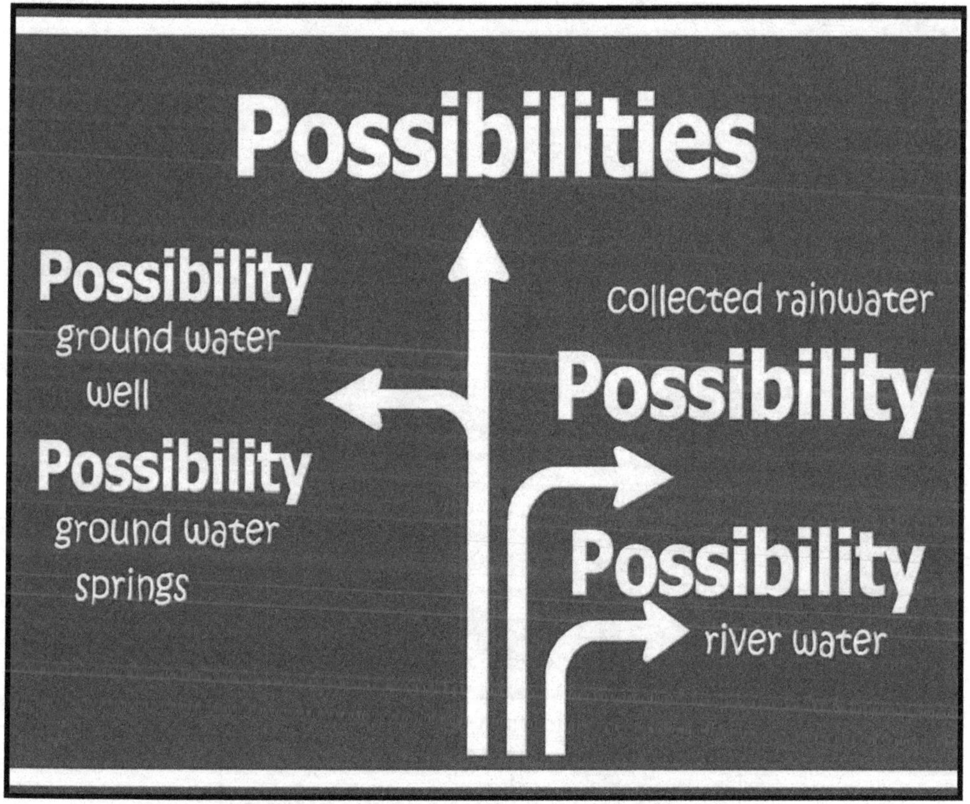

"Given a choice ..."

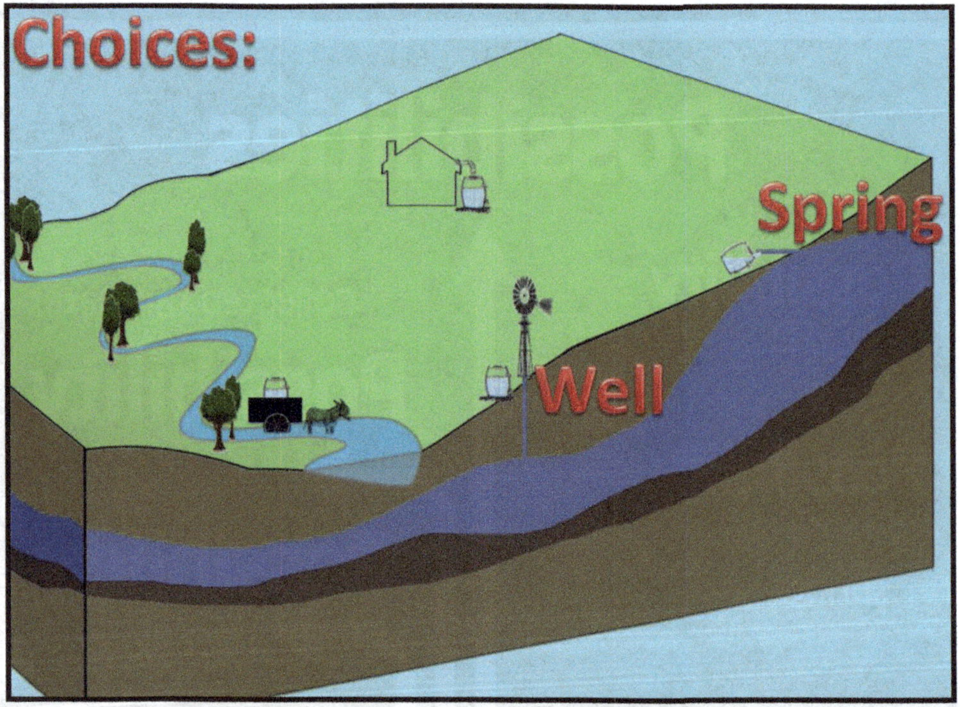

"...people mostly preferred groundwater (well or spring)..."

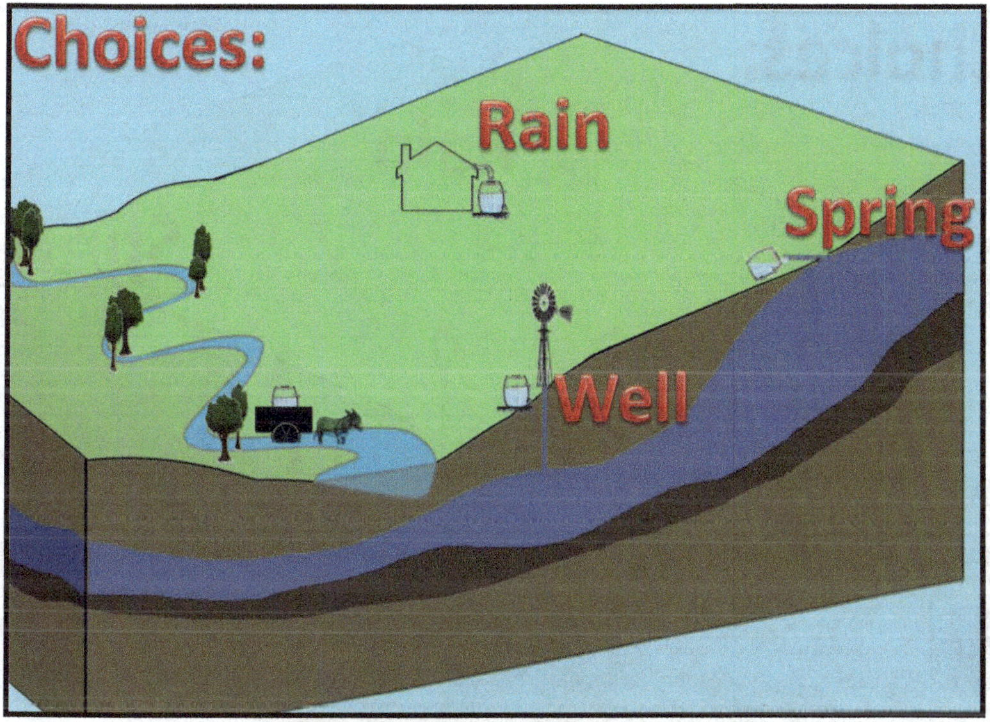

"...secondly, rain water (when it eventually rained)..."

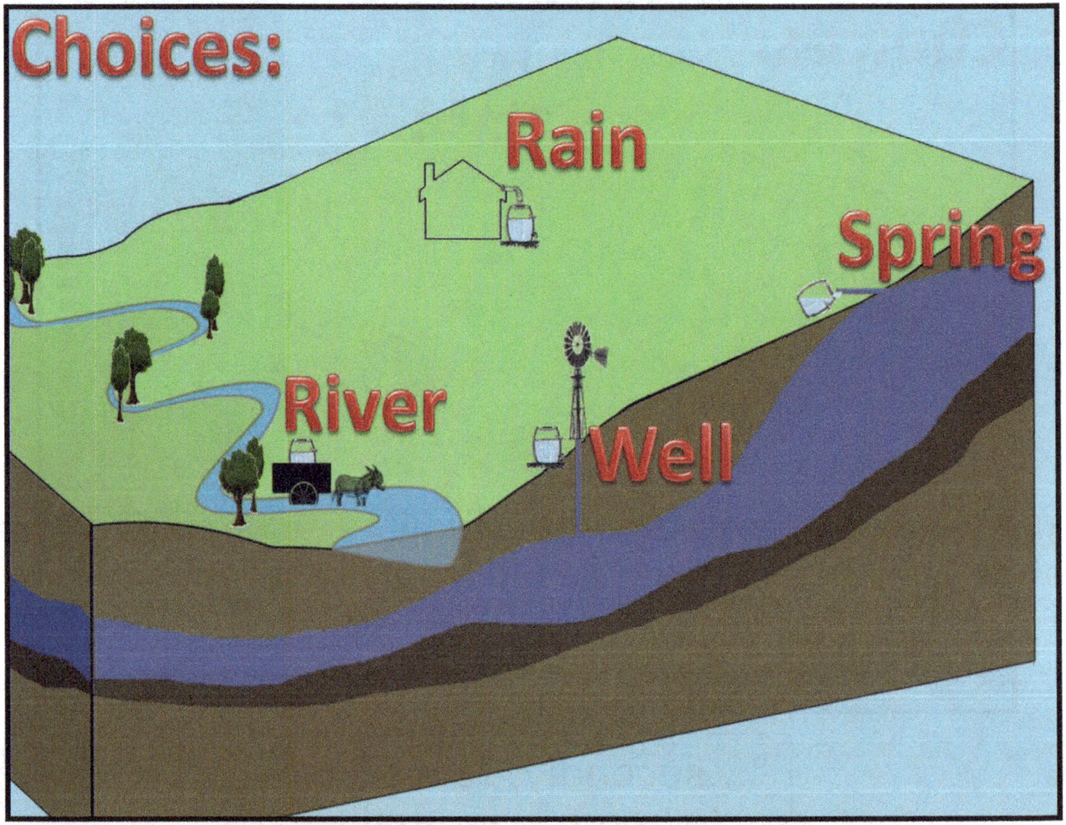

"...and finally, river water."

"There weren't many choices."

"They felt lucky to have any water at all."

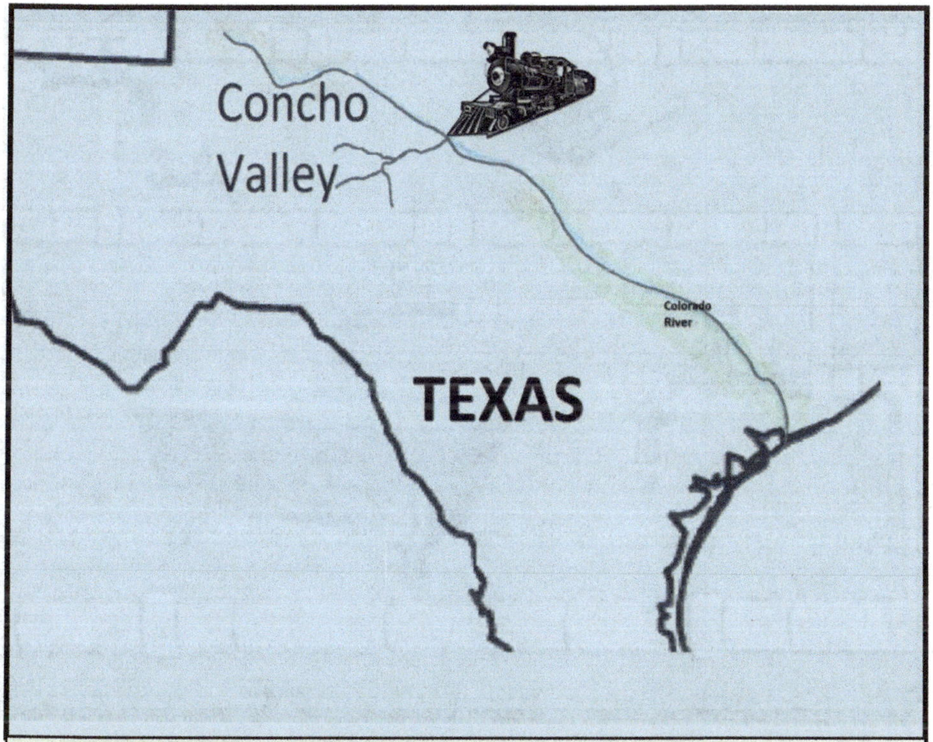

"As time went on, people from all over the world found that they could make a living, here in the San Angelo area."

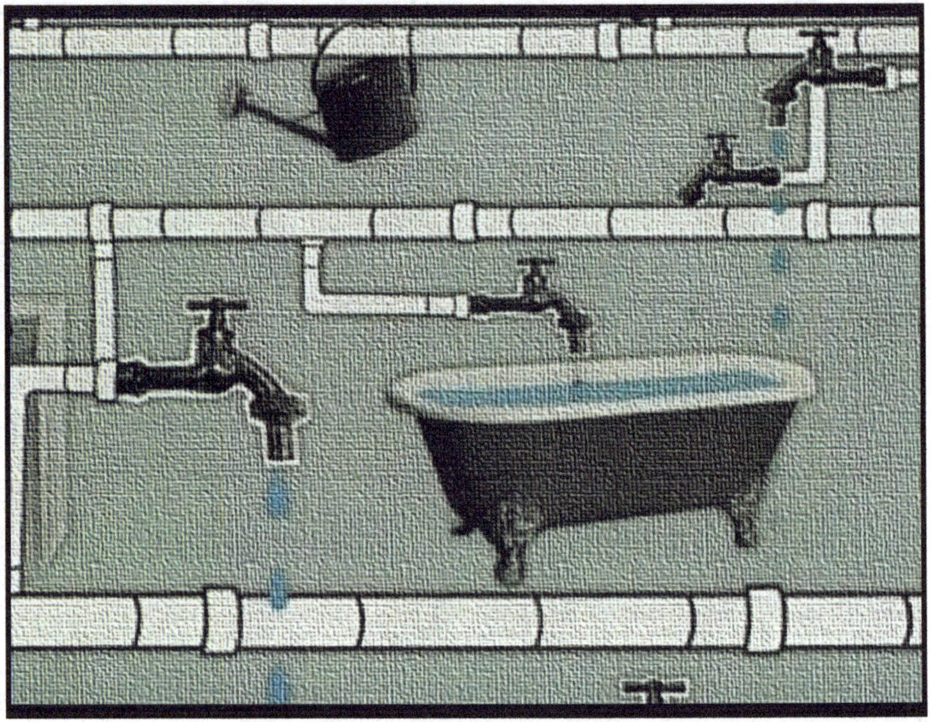

"And when the steam pump became available in San Angelo, water was pumped out of the river and easily piped to homes and businesses."

"My grassland became 'people-inhabited.' The people built streets, and all kinds of buildings."

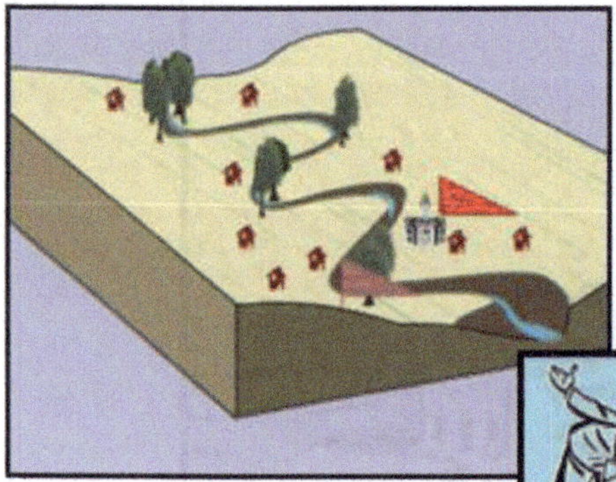

Later in time

It's OK to pump a lot of water!

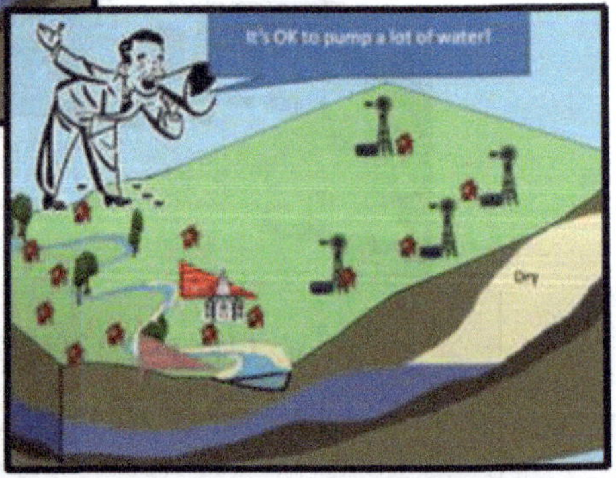
It's OK to pump a lot of water!
Dry

It's OK to pump a lot of water!
Dry

Later in time

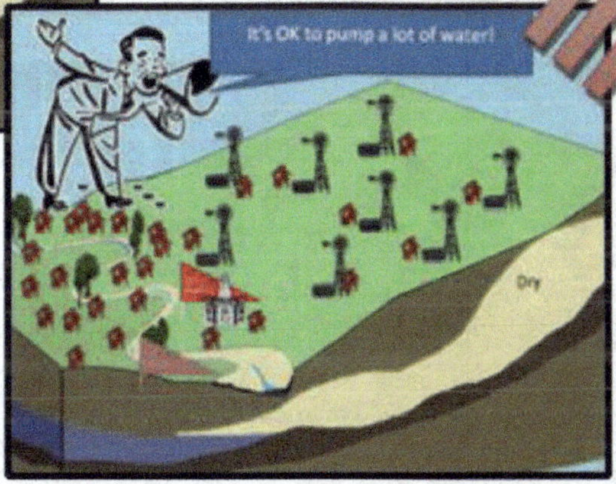
It's OK to pump a lot of water!
Dry

"And because water was much easier to get, the people began to use more of it."

"They planted and watered trees for beauty and comfort from the sun."

"Ranchers raised and watered herds of fenced-in animals."

"Farmers plowed the land and began to plant and water crops."

Mesquites steal water.

“Sadly, during all this progress, mesquite trees became more common than grasses.”

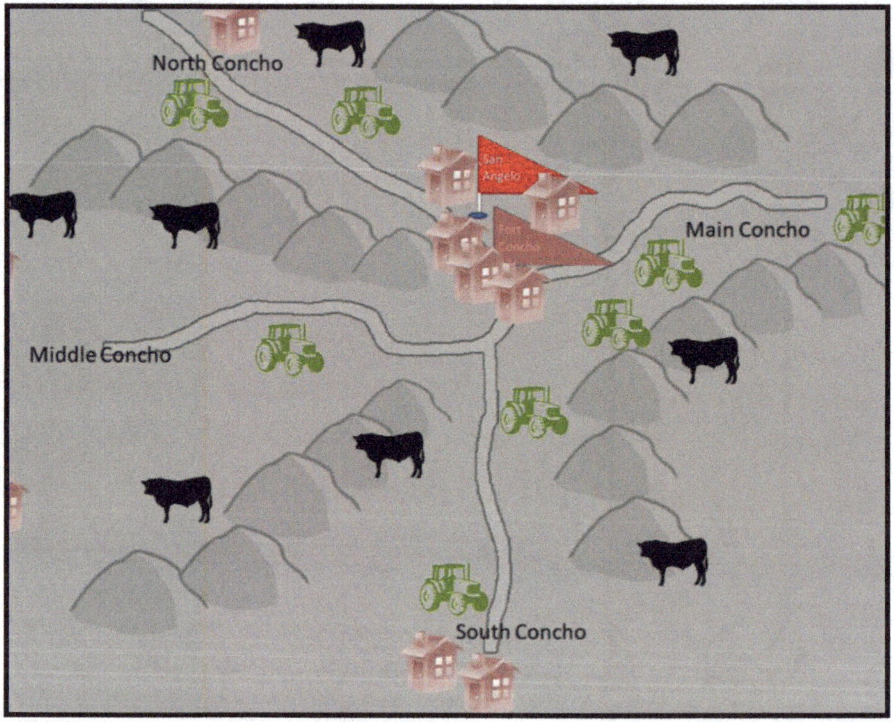

"And so my river's grassland changed all around me and this affected my river," she said.

Chapter 4
Dams

Dezzerae made ripples in the water with
a long stick she had found.
The ripples of the still water
made her sleepy.

The mermaid changed the subject.
"Do either of you know what a dam is?"

Haley said proudly, "Beavers block a river with wood to make a dam."

"Very good! Beavers make a natural dam, but people also make dams."

"Dirt, stone, wood or concrete dams were made along the river and its creeks for water storage."

"Dammed river irrigated crops or watered livestock or was piped to the entire growing city," she said.

"But, dams can be a problem," she continued.
"They slow the flow of the water."

"Slow and shallow: means stinky,"
Dezzerae remembered.

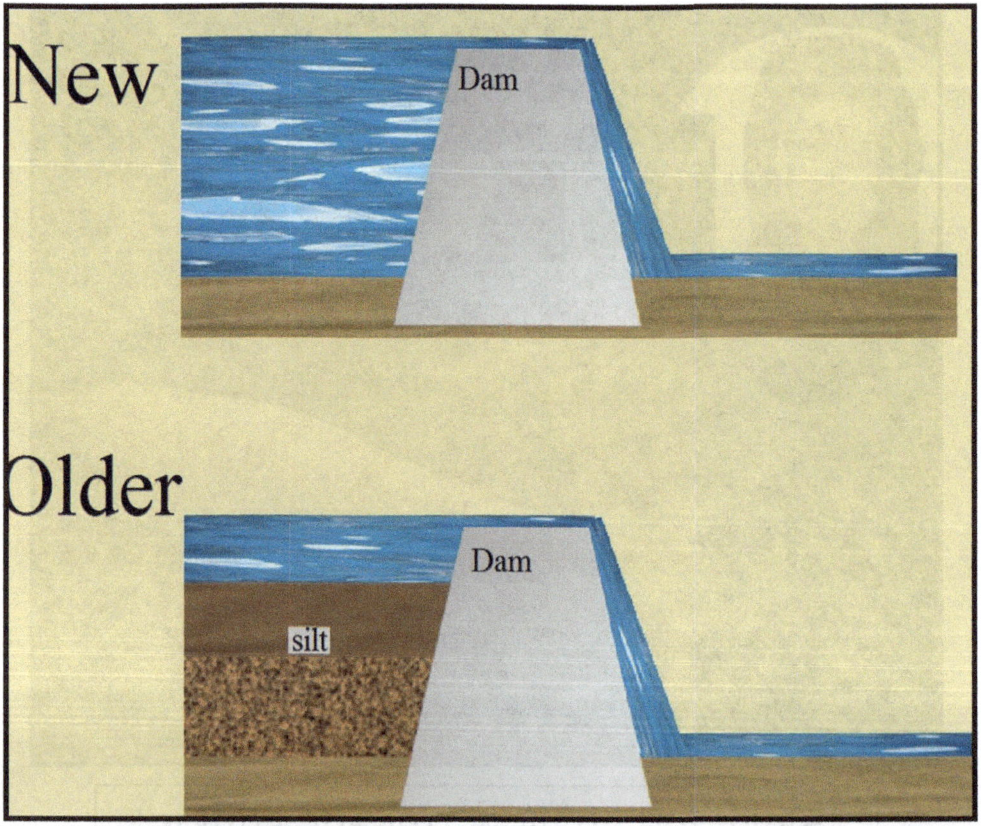

"Yes, and dams collect silt."

“Too much silt makes the river shallow.”

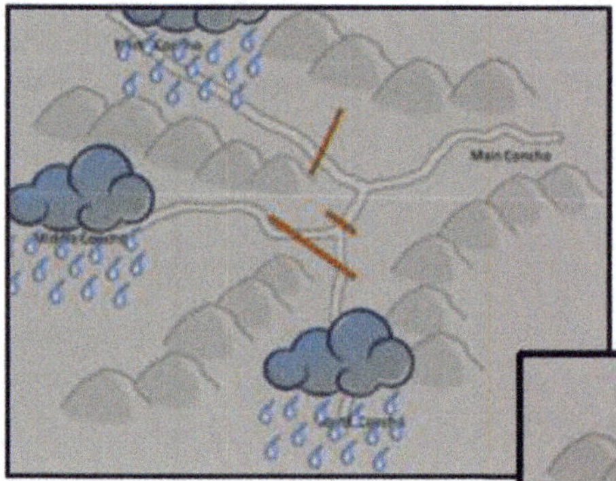

Later in time

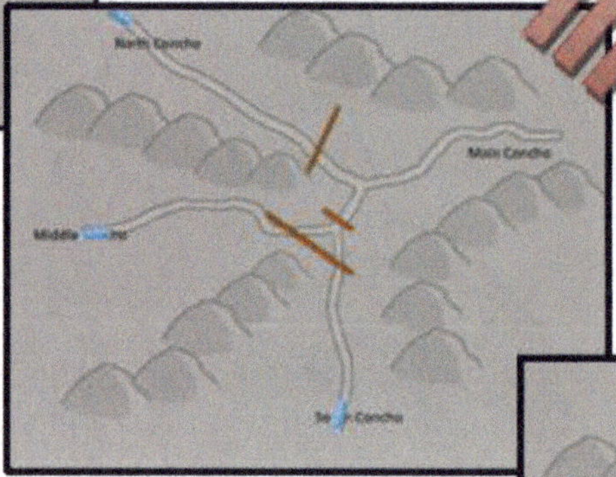

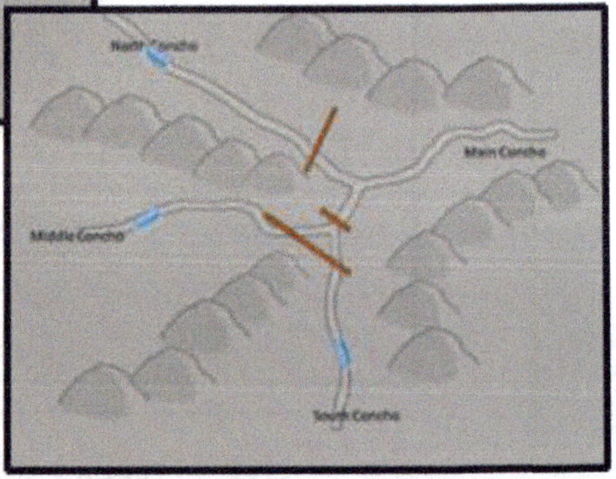

Three Dams

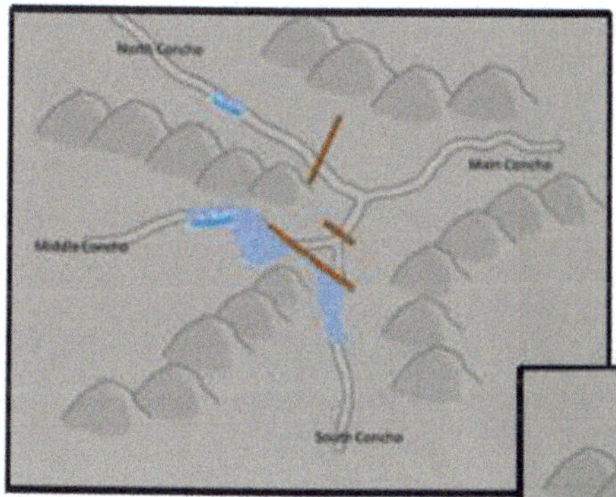

Later in time

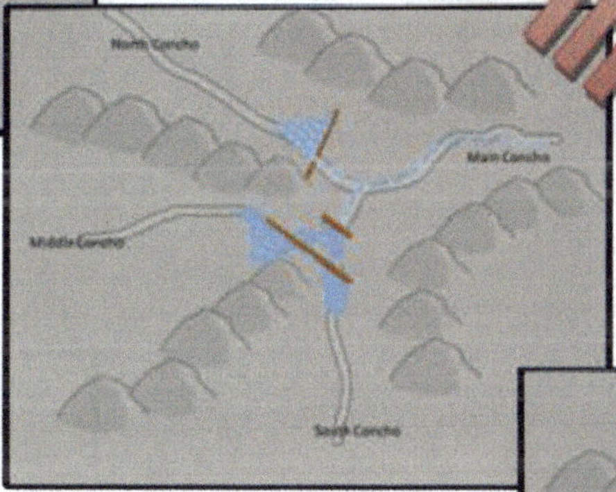

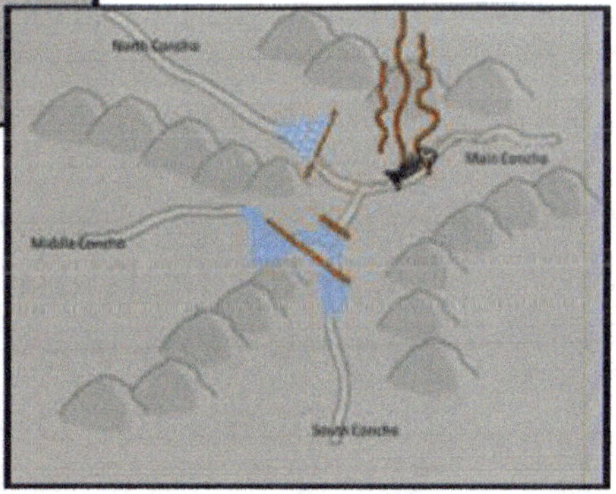

"Dams prevent 'cleansing rains' from cleaning the river."

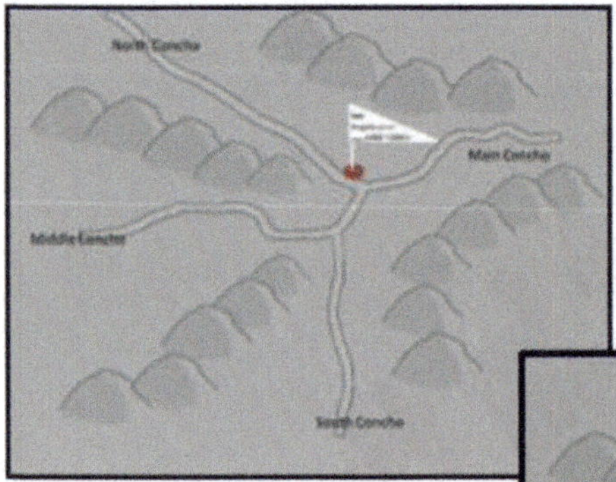

Later in time

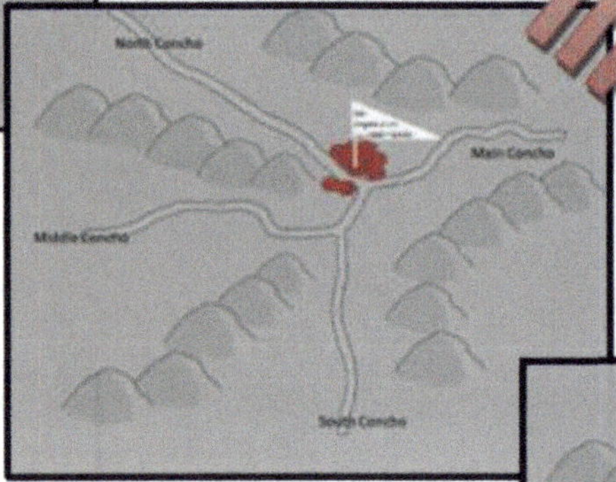

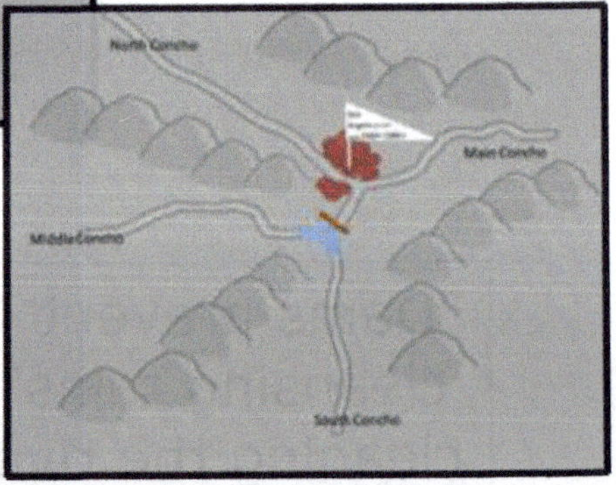

"As the human population went up, the need for dams…"

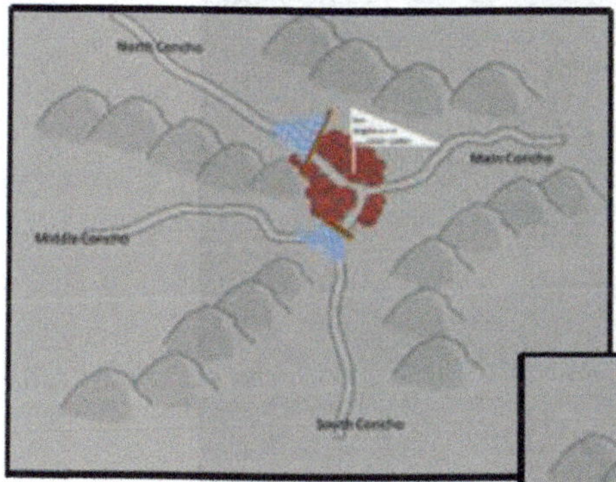

Later in time

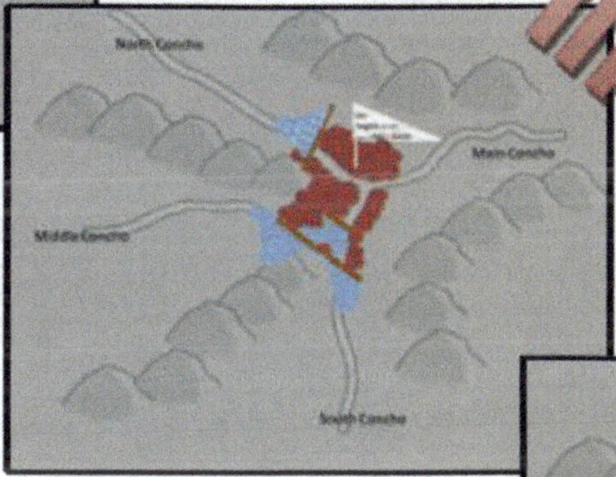

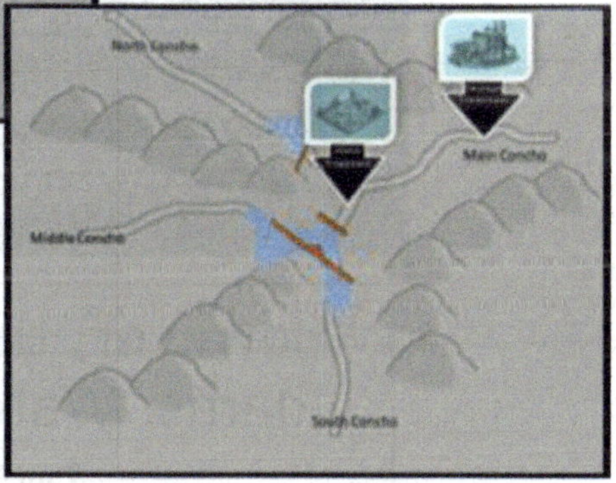

"...water treatment and sewage disposal occurred," she said.

"However, there are many benefits from dams. Water storage and beauty are two," she continued.

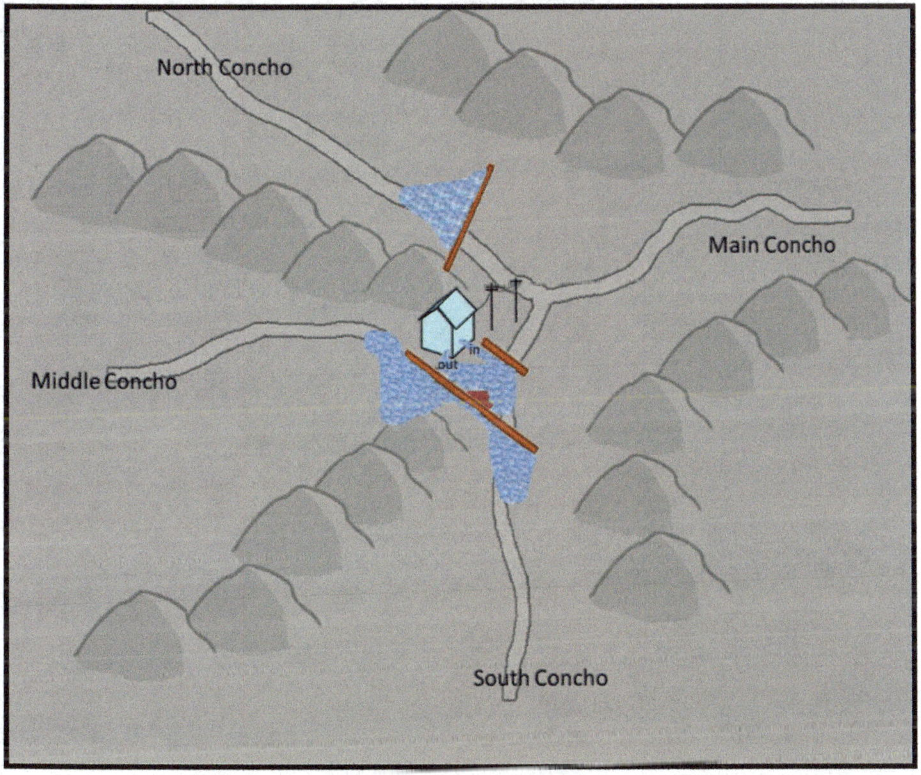

"Some dams
can help electric power plants."

"In San Angelo, dammed water cooled the steam of the electric generators."

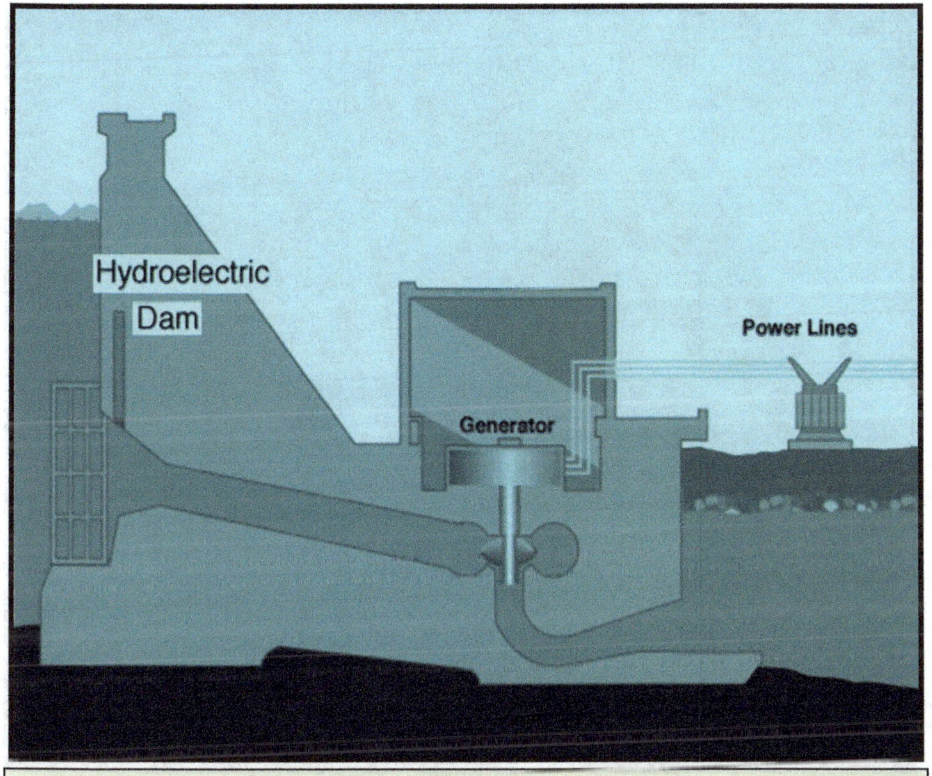

"In other places, moving dammed water can actually create electricity."

"Finally, if a dam is big enough, it can prevent 'killing and destroying' floods."

Photo: West Texas Collection, ASU 1906 Flood, Oakes facing north

"I lived through three really bad floods in 1882, 1906, and 1936."

Photo: West Texas Collection, ASU
1936 Flood, view from Cactus Hotel

I936 Flood

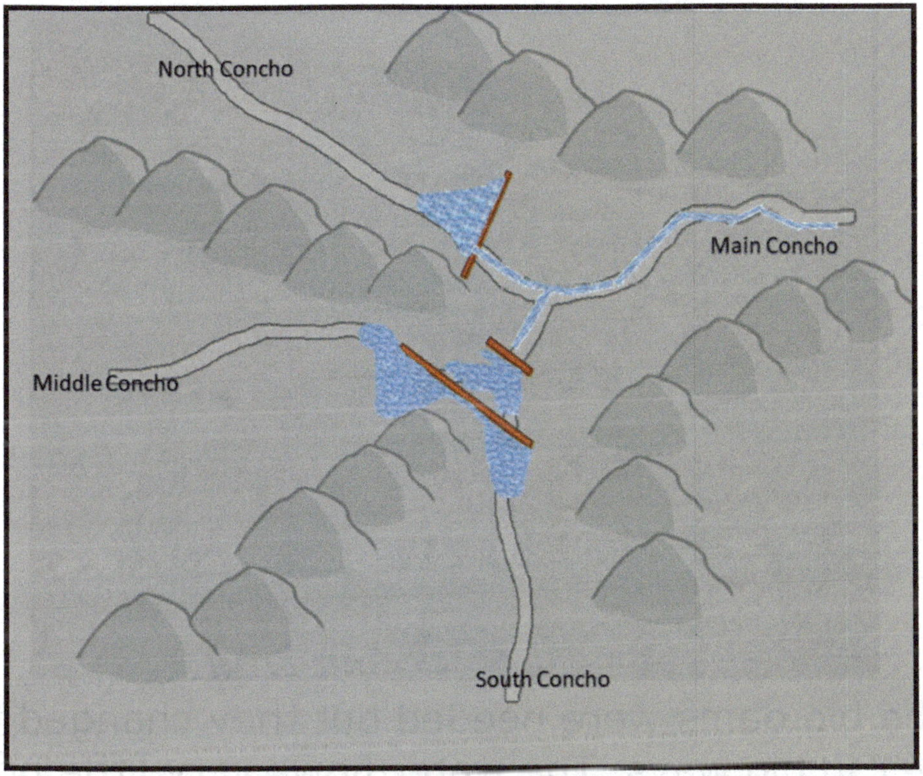

"San Angelo has three large dams now and doesn't worry about flooding."

"The big dams were needed but they changed the river in big ways. The water flows very little now. And 'cleansing rains' just don't have a chance to cleanse, anymore."

Chapter 5
On The Ground

Two guys skated past the kids who, by now, were wondering why no else seemed to notice the talking mermaid.

Finally, Dezzerae had the courage to ask, "How can a statue talk? Why are you telling us this story?"

She replied, "Everyone needs to know the river's story because the river is in a bit of trouble now."

"So many people like to live in San Angelo and all of them, even you kids sometimes hurt the river."

"Cars on streets drip toxic chemicals which flow into the river during a rain."

"People put fertilizers and chemicals on their yard and much of it ends up in the river because of rain."

Surprised, the children said,
"But we're not hurting the river!"

She continued, "Some people let their garbage blow away."

"When rains come or when winds blow, all kinds of dirt..."

"...leaves, chemicals ..."

"...and trash end up in the river."

"Most people don't realize that they pollute the river by just living in the city."

"And sadly, the pollution stays here because of the dams."

Haley had two solutions to the problem.
"If dams are the problem,
then get rid of them."

"If there are too many people here, then make some of them move."

But she replied, “If we remove the dams, there could be water shortages ...”

"...or possibly floods."

"And, Haley, who should be forced to move away from this city: you kids and your families?"

Everything went quiet.
They thought and thought.
Soft raindrops began to fall.

One by one, the butterflies left the mermaid for shelter.

Soon only one was left.

With the sun still shining, the sprinkle of raindrops would most likely stop soon. West Texas rains are like that.

But Miranda’s dad closed the book
and searched the sky.
Would it rain or not?

A tear rolled down the mermaid's cheek
as she softly concluded,
"Take care of my living river."

"If you put something on the ground, it will probably end up in my waters."

Chapter 6
Whose River?

A strong gust of wind blew the remaining butterfly away. The mermaid spoke no more.

Events of the Celebration Bridge became more noticeable to the children.

A dog barked on the sidewalk ahead.

Pigeons flew down from the tall bridge.

They realized that the story had ended and that a storm was approaching.

But just then, a crawdad on a bamboo bloom grabbed Dezzerae's stick with its claw.

At the same time, Miranda’s dad decided,
“It looks like rain.”
“But look Dad. Can we keep it?”
Miranda asked while pointing at the crawdad.

He hesitated for a moment, as parents often do, but finally decided, "No, let's leave it in its natural home, and we need to go home, too."

So they let the crawdad go back into its home, the river; and they left for their homes in the city, by the river.

The children told the mermaid's story to their parents, families and friends.

Everyone wondered how they'd learned such an important story, because no one believed that the mermaid actually spoke to them.

Yet, everyone did agree that preventing litter and pollution would not be that hard to do.

During later visits, the children tried again and again to get the mermaid to speak.

But the magic of the butterflies was never there again.

When you pass the welcoming mermaid of the Celebration Bridge, especially during the autumn monarch butterfly migration, see if she will speak to you, too.

But please, do what you can every day to prevent trash and debris from getting on the ground.

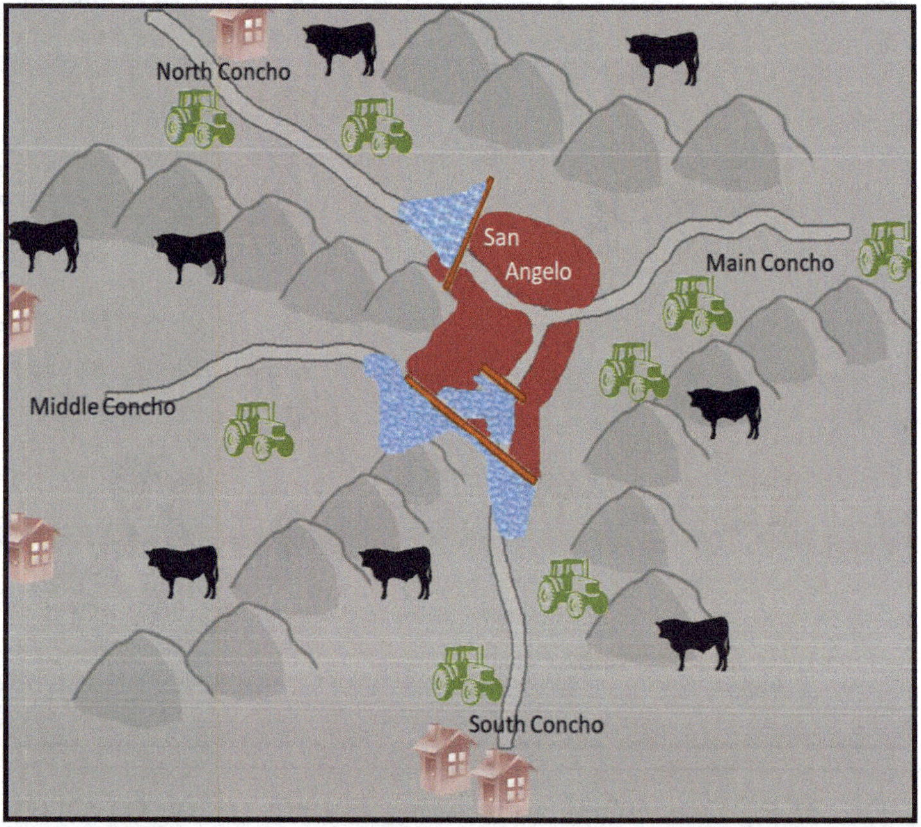

Progress changed the land and the river.

Human ingenuity can and will find a way for the city and its river to coexist beautifully.

She thought she was the keeper of the river,
but WE are the keepers of the river.

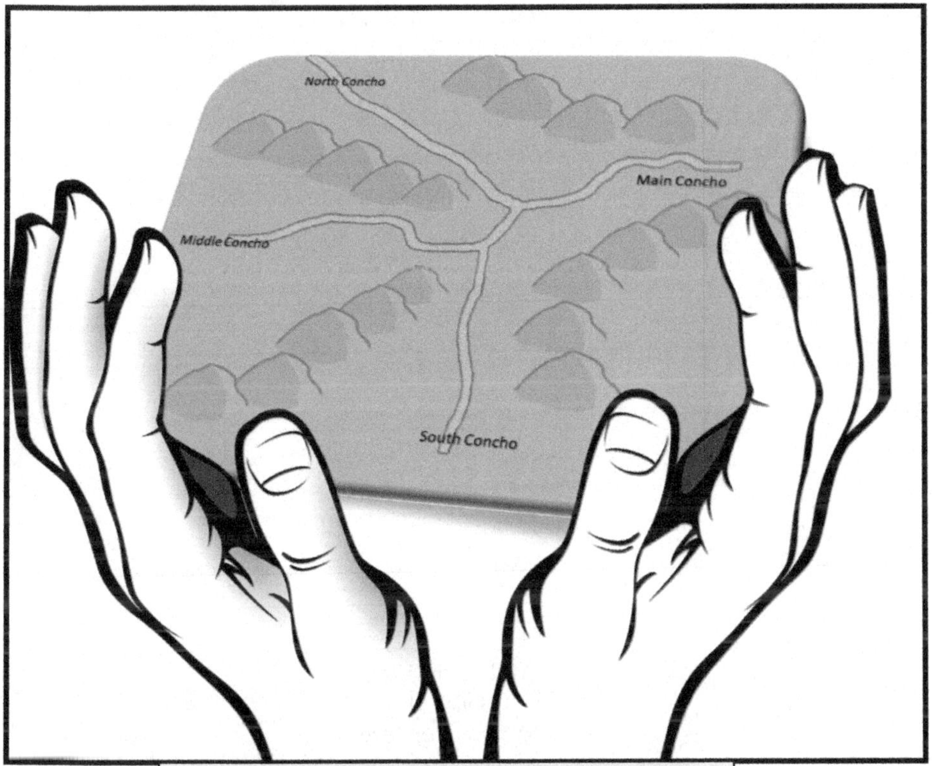

She spoke of OUR River.

Made in the USA
Monee, IL
02 November 2026

67387478R00103